Thilo L. Schenck

Role of AMP Kinase in Volume Control of Colonic Crypts

Thilo L. Schenck

Role of AMP Kinase in Volume Control of Colonic Crypts

AMP-activated Kinase controls Regulatory Volume Increase of Colonic Crypts by modulating CFTR and NHE1 activity

Südwestdeutscher Verlag für Hochschulschriften

Impressum/Imprint (nur für Deutschland/only for Germany)
Bibliografische Information der Deutschen Nationalbibliothek: Die Deutsche Nationalbibliothek verzeichnet diese Publikation in der Deutschen Nationalbibliografie; detaillierte bibliografische Daten sind im Internet über http://dnb.d-nb.de abrufbar.

Alle in diesem Buch genannten Marken und Produktnamen unterliegen warenzeichen-, marken- oder patentrechtlichem Schutz bzw. sind Warenzeichen oder eingetragene Warenzeichen der jeweiligen Inhaber. Die Wiedergabe von Marken, Produktnamen, Gebrauchsnamen, Handelsnamen, Warenbezeichnungen u.s.w. in diesem Werk berechtigt auch ohne besondere Kennzeichnung nicht zu der Annahme, dass solche Namen im Sinne der Warenzeichen- und Markenschutzgesetzgebung als frei zu betrachten wären und daher von jedermann benutzt werden dürften.

Coverbild: www.ingimage.com

Verlag: Südwestdeutscher Verlag für Hochschulschriften GmbH & Co. KG
Heinrich-Böcking-Str. 6-8, 66121 Saarbrücken, Deutschland
Telefon +49 681 37 20 271-1, Telefax +49 681 37 20 271-0
Email: info@svh-verlag.de

Herstellung in Deutschland (siehe letzte Seite)
ISBN: 978-3-8381-3179-5

Imprint (only for USA, GB)
Bibliographic information published by the Deutsche Nationalbibliothek: The Deutsche Nationalbibliothek lists this publication in the Deutsche Nationalbibliografie; detailed bibliographic data are available in the Internet at http://dnb.d-nb.de.

Any brand names and product names mentioned in this book are subject to trademark, brand or patent protection and are trademarks or registered trademarks of their respective holders. The use of brand names, product names, common names, trade names, product descriptions etc. even without a particular marking in this works is in no way to be construed to mean that such names may be regarded as unrestricted in respect of trademark and brand protection legislation and could thus be used by anyone.

Cover image: www.ingimage.com

Publisher: Südwestdeutscher Verlag für Hochschulschriften GmbH & Co. KG
Heinrich-Böcking-Str. 6-8, 66121 Saarbrücken, Germany
Phone +49 681 37 20 271-1, Fax +49 681 37 20 271-0
Email: info@svh-verlag.de

Printed in the U.S.A.
Printed in the U.K. by (see last page)
ISBN: 978-3-8381-3179-5

Copyright © 2012 by the author and Südwestdeutscher Verlag für Hochschulschriften GmbH & Co. KG and licensors
All rights reserved. Saarbrücken 2012

TABLE OF CONTENTS

I.	German abstract	1
II.	English abstract	2
III.	Introduction	3
IV.	Literature review	4
	COLON	4
	RVI	8
	AMPK	9
V.	Hypothesis	9
VI.	Material and Methods	10
	Laboratory Animals	10
	Solutions and substances	10
	Methods	12
	Statistical analysis	20
VII.	Results	21
	Amiloride shows participation of NHE in RVI of colonic crypt cells (Protocol 1)	22
	Compound C suggests effect of AMP Kinase on NHE during RVI (Protocol 2)	25
	AMP Kinase-dependent NHE regulation is specific for volume changes (Protocol 3)	26
	Effect of CFTR on RVI (Protocol 4)	27
	Changes of Calcium concentrations during cell recovery (Protocol 5)	28
	AMP Kinase and CFTR (Protocol 6)	29
VIII.	Discussion	30
IX.	Conclusions	31
X.	List of Abbreviations	32
XI.	References	33
	Figures	33
	Literature	34

I. GERMAN ABSTRACT

Wirkung der AMP aktivierten Kinase auf die Aktivität von CFTR und NHE1 zur regulatorischen Erhöhung des Zellvolumens (RVI) in Kolonkrypten, unter besonderer Berücksichtigung der AMP-Kinase.

Einleitung: Der Darm ist das wichtigste Organ für Absorption bzw. Sekretion von Nahrung, Flüssigkeit und Elektrolyten. Im Kolon ist die Sekretion von Salz und Flüssigkeit an die Aktivität des apikalen CFTR-Proteins und des basolateralen NKCC-1 Transporters gekoppelt. Aktiviert von ATP, können diese Membranproteine zu einem Transport großer Mengen von Flüssigkeit und Elektrolyten durch die Zelle führen. Es ist ein Mechanismus anzunehmen, der starke Schwankungen des Zellvolumens während dieser transzellulären Transportvorgänge verhindert. Eine Beteiligung von NHE1 an der regulatorischen Erhöhung des Zellvolumens (RVI) ist bekannt. Die hierfür verantwortliche Regulation von NHE1 ist jedoch noch unbekannt.

Methoden: Kryptenzellen des distalen Rattenkolons wurden einem osmotischen Schock (von 295 mOsm auf 344 mOsm) ausgesetzt, um eine rasche Änderung des Zellvolumens zu provozieren. Zur Beobachtung des transmembranösen Transportes von Chlorid (durch CFTR), Protonen (durch NHE1) und Kalzium wurden Fluoreszenzfarbstoffe und eine in Echtzeit laufende Hochgeschwindigkeitskamera verwendet.

Ergebnisse: Die Inhibiton von NHE1 durch Amiloride nimmt den Kolonzellen die Möglichkeit zur RVI. Die AMP-Kinase hemmende Substanz Compound C reduziert die Aktivität von NHE1 während der RVI, hatte jedoch keinen Effekt auf die Reaktion der Zelle auf eine intrazelluläre Ansäuerung. Die AMP-Kinase aktivierende Substanz AICAR konnte die Aktivität von CFTR vermindern, hatte aber keinen Effekt auf die Aktivität von NHE1. Eine Hemmung des CFTR durch NPPB scheint die RVI nicht zu beeinflussen. Während der RVI wurden keine nennenswerten Änderungen der intrazellulären Kalziumkonzentration beobachtet.

Schlussfolgerung: Diese Arbeit beschreibt eine neue Funktion der AMP aktivierten Kinase und zwar die eines Volumensensors, der das apikale CFTR-Protein und den basolateralen NHE1 reguliert. Dadurch kann die AMP aktivierte Kinase den apikalen Austritt und den basolateralen Eintritt von Kochsalz so modulieren, dass ein konstantes Zellvolumen beibehalten wird. Diese Regulation des Elektrolyttransports durch die AMP aktivierte Kinase ist spezifisch für die Zellvolumenregulation und hat keine Funktion für die Aufrechterhaltung eines konstanten intrazellulären pH. Somit könnte die AMP aktivierte Kinase der gesuchte Regulator von NHE1 sein.

II. ENGLISH ABSTRACT

Role of AMP-activated Kinase in controlling Regulatory Volume Increase of Colonic Crypts by modulating CFTR and NHE1 activity with special emphasis on the aspect of AMP-activated Kinase.

Background: The intestines are the primary site for nutrient, fluid and electrolyte absorption and secretion. Fluid and salt secretion in the colon are linked to activation of the apical CFTR protein and the simultaneous activation of the basolateral NKCC-1 transporter. Activated through ATP generation these proteins can lead to massive transportation of fluid and electrolytes across the cell. During this large transcellular movement of ions, a mechanism must be in place to prevent drastic shifts in cell volume. NHE1 is known to be involved in regulatory volume increase (RVI) but its regulator still remains unknown.

Methods: By applying an osmotic shock (from 295 mOsm to 344 mOsm) a rapid change in cell volume was caused in rat colonic crypt cells. High speed real time video imaging was used to monitor the resulting transmembranous movement of Cl^- (CFTR), protons (NHE1) or calcium. Furthermore, the cells' response to intracellular acidification was observed.

Results: Inhibiting NHE1 with Amiloride removes the cell's ability to perform RVI. Exposure to the AMP Kinase inhibitor Compound C resulted in reduction of NHE1 activity in RVI but did not change the cell's response to acidification. Exposure to the AMP Kinase activating agent AICAR resulted in a reduction in CFTR activity, while having no effect on NHE1 activity. Inhibiting CFTR with NPPB does not seem to influence RVI. During RVI no major changes of intracellular Ca^{2+} were observed.

Conclusion: The present study describes the role of AMP Kinase acting as a "volume sensor" which regulates apical CFTR and basolateral NHE1 activity. Thereby AMP Kinase modulates apical efflux and basolateral entry of salt specifically to maintain a constant cell volume, while having no effect on intracellular pH maintenance, and might be the upstream regulator of NHE1.

III. INTRODUCTION

Diarrhea still accounts for millions of deaths – directly or by participation – each year. Cholera, a life threatening disease, causes massive loss of electrolytes and water through the gastrointestinal tract and still kills more than 10.000 individuals every year (WHO statistics). With proper rehydration treatment, like the oral rehydration therapy (ORT) promoted by the WHO, death can be prevented in many cases but it is not a cure for the disease nor does it stop the outflow of water and the loss of electrolytes, thus mothers in the Third World sometimes stop giving the life-saving ORT to their children because they see the diarrheas worsening as a sign of fluid regain.

Up to date, we know that the cystic fibrosis transmembrane receptor (CFTR), a major chloride secreting channel, is responsible for the massive fluid loss in cholera and other secretory diarrheas. The cAMP-activated pathway of CFTR, also activated by cholera toxin, has been studied thoroughly as well as the cell's reaction to shrinkage, the process following every loss of electrolytes and water. Still, no proper treatment to stop the secretion of chloride and thereby resolving the cause of the diarrhea could be established. If we were to find another pathway to regulate CFTR activity, we might be able to inhibit CFTR selectively and thus stop secretory diarrhea. Assuming there is a pathway, what other functions might it have in the regulation of cell volume? Would its blockade threaten other compartments?

In heart research, scientists have found an enzyme that stabilizes cell membrane potential and might be crucial to the development of arrhythmias after an ischemic heart attack. This enzyme is termed the "AMP-activated Kinase" (AMPK). It starts to phosphorylate and therefore inactivate energy consuming processes after being activated by the enhancement of the AMP:ATP ratio, which is a clear sign that the cell lacks energy. Maybe this effect cannot only be seen in the heart but also in the colon, where a loss of chloride means a loss of energy to the cell (and therefore the entire body). Could AMPK be an inhibitory pathway of CFTR?

And if so, could AMPK act as a volume sensor in the cell? It is known that sodium-hydrogen-exchanger isoform 1 (NHE1) and NKCC are the major electrolyte importers of the cell in volume increase. Might AMPK also be able to influence them? For years, research groups have been trying to find the sensor of the activating pathway of NHE1. Could AMPK be this sensor?

The present study tries to answer the questions above. Using vital colonic crypt cells, a high speed video fluorescence microscope to visibly observe changes in cell volume, and pH, calcium and chloride sensing dyes, the activity of CFTR and NHE can be measured. Cells can be observed under altered activity of AMPK by using the AMPK activating and inhibiting agents AICAR and Compound C.

IV. LITERATURE REVIEW

COLON

Gross anatomy

The colon extends from the distal end of the ileum to the anus, to a total length of approximately 1.5 m. Together with cecum, appendix, rectum and anal canal it is known as the large intestine. In gross anatomy four major parts of the colon can be differentiated by localization: the ascending, the transverse, the descending and the sigmoid colon. Functionally we differentiate between a proximal and a distal colon with its border at the height of the left colic flexure. At this site, arterial blood flow switches from the upper mesenteric artery to the lower mesenteric artery, connected through the marginal artery. The venous supply drains entirely into the portal vein. Via pre-aortic lymph nodes, lymphatic drainage of the colon enters the cisterna chyli. Nerval innervation consists of the sympathic, the parasympathic and the enteric nervous system[1].

Histology

Like most parts of the intestine the colon consists of several layers, from the luminal side to the peritoneal cavity: mucosa (epithelium, lamina propria and muscularis mucosa), submucosa, muscularis externa and serosa. The mucosa is built up by numerous straight tubular glands which extend through the full thickness of the mucosa. Since luminal microscopic pictures reveal a plain surface with orderly patterned openings, the glands are being referred as crypts. These crypts consist of columnar epithelium, which is built up by enterocytes (sometimes also referred as colonocytes). They arise from stem cells at the bottom of the crypt and move to the top where they are being dismissed after about six days. Far less numerous than absorptive enterocytes, mucus producing goblet cells and neuroendocrine cells are found in the colonic crypt[2,3].

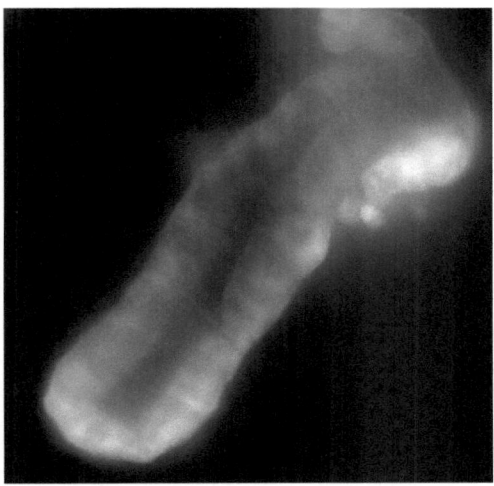

Figure 1: A colonic crypt separated by the EDTA method and coloured by BCECF in 400x magnification (Hauser & Schenck 2007).

Physiology

The main function of the colon is to resorb water and electrolytes and to eliminate all undigested waste. Up to 1.8 L of electrolyte rich fluid are being absorbed in the human colon every day[4], but fluid uptake and secretion vary to maintain normal stool consistency and final stool ion concentrations of <5 mM Na^+, 2 mM Cl^-, and 9 mM K^+ [5]. Geibel and Binder could show in 2001 that fluid absorption is mainly Na^+-dependent. Substituting a lack of Na^+ and Cl^- by NMDG results in a greatly reduced net fluid absorption[6].

Goblet cells produce mucine containing mucus to "grease" the inner surface of the colon, in order to keep the stool from sticking to the epithelium.

Stating R. Greger, the following ion transport mechanisms have been described: electroneutral absorption of Na^+ and Cl^- via the parallel arrangement of Na^+/H^+ and Cl^-/HCO_3^- exchangers, rheogenic Na^+ absorption via epithelial Na^+ channels (ENaC), absorption of K^+ via the H^+/K^+-ATPase located in the luminal membrane, absorption of short fatty acids mostly by the uptake of the anion, K^+ secretion via K^+ channels in the luminal membrane and Cl^- secretion via luminal membrane Cl^- channels[7].

Surface ion transporters

Since surface membrane ion transporters and channels play a vital role to this thesis, they are described in detail below:

NHE

The sodium-hydrogen-exchanger (NHE) is a transmembrane exchanger, found on the cell surface of every cell in the mammalian organism. Its main function is to pick up an extracellular sodium-ion to exchange it for an intracellular hydrogen-ion[8,9]. Thus, the electroneutrally working transporter has an effect on salt concentration and pH in and around each cell.

Up to date, the NHE gene family consists of nine expressed isoforms. They share a distinct homologue structure consisting of a transmembrane domain with 12 membrane-spanning helices and a big intracellular domain. The transmembrane region is found to be only slightly modified throughout the isoforms and is responsible for sensitivity to drugs like Amiloride. The intracellular C-terminal domain shows more variations which explain the different activity in each isoform[10,11].

Among the eight other known isoforms of its family[12,13,14,15,16,17,18,19,20,21,22], the ubiquitous plasma membrane exchanger NHE1 has been shown to possess an important function as a cell volume regulator, pumping osmotic active sodium into the cell in exchange for an equal amount of hydrogen ions which are retrieved through the HCO_3^- buffer system. The osmotically active sodium ion drags water into the cell through aquaporin channels like aquaporin 1. This effect has been thoroughly studied in a variety of organisms (e.g. Amphiuma, dog red blood cells, nectures gallbladder cells, mammalian lymphocytes, chinese hamster ovary cells), especially during regulatory volume increase (RVI, see next chapter)[23,24,25,26,27,28].

Several mechanisms have been suggested to explain the induction of RVI mediated NHE1 activity: allosteric regulation, G-protein coupled activation and a calmodulin associated activated Jak2 mechanism. Yet, none of these mechanisms have been proven so far[29]. Still the Jak2 induced calmodulin coupling to NHE1 appears to be most promising, but the existence of an upstream activator remains unknown[30]. Ezrin, a protein that regulates the distribution of NHE1, has been shown to be phosphorylated during cell shrinkage, but it is unclear if this effect precedes or follows NHE1 activation. On the other hand, it seems to be evident that NHE1 phosphorylation does not significantly change during RVI[31,32,33,34]. Krump and his group could show that NHE1 activation is specifically induced by cell shrinkage and not by hyperosmolarity[35]. Furthermore, Fuster et. al. were able to prove that NHE1 activity is not increased by mechanical alteration of cell volume[36].

Beside its volume regulatory function, NHE1 is also known as the housekeeper[37,38], but contrary to NHE2 and NHE3 it is not upregulated by Na^+ depletion[21].

The Na^+/H^+ exchanger isoform 2 is the predominant NHE isoform in murine colonic crypts and its lack causes NHE3 upregulation. Beside that, its function remains obscure[39].

NHE3 is expressed on the apical plasma membrane of renal and intestinal epithelium[40,41,42,43] and seems to be important to intravascular volume maintenance and therefore blood pressure [44,45].

NHE6-9 are less well described and beside NHE8 located mainly intracellularly. Their function still remains to be discovered and might even be different from the exchange of sodium and hydrogen ions[46,47,48].

CFTR

The apical cystic fibrosis transmembrane receptor (CFTR) is the major Cl⁻ secreting channel in secreting epithelial cells and its name derives from its deficiency in cystic fibrosis (CF). Structurally, CFTR is a member of the ATP-binding cassette[49] and is found to be activated by a variety of pathways including cAMP, Ca^{2+}/calmodulin dependent kinases, PKA, PKC and cGMP-activation[50]. A link between CFTR and AMPK has previously been shown for the T84 cell line – a colon derived cancer tissue, but physiological data remain to be collected[51]. CFTR is responsible for both physiologic and pathologic chloride secretion with a close link to a variety of other channels and transporters of chloride and sodium, e.g. NKCC-1, ENaC and NHE1[3,52,53,54]. The importance of CFTR gets obvious when pathologic changes in the protein or its activating pathways occur. Cystic fibrosis is an inherited disease caused by an autosomal recessive mutation of the CFTR channel, leading to decreased pulmonary mucus production, sweat irregularities, impaired pancreatic enzyme function and chronic constipation etc. In cholera, the choleratoxin causes an increase of intracellular cAMP. This leads to permanent activation of the CFTR channel and causes life-threatening diarrhea with fluid excretion rates up to 6 L per hour[55]. Thus, finding a way to regulate CFTR function remains to be an important task of medical research.

NKCC

The sodium-potassium-2-chloride transporter has been best studied in the kidney. By taking up four extracellular ions at a time, NKCC is a potent salt absorber and therefore important for water absorption. Thus, beside NHE, it is important to RVI. This can be seen by withdrawing potassium or using the inhibiting agents furosemide or bumetanide which show a significant change in RVI [56,57].

ENaC

The electrogenic sodium channel (ENaC) is expressed on the apical membrane of epithelial cells. EnaC has absorptive functions and is sensitive to Amiloride. It is highly selective for Na^+ and allows Na^+ flow only into the cell. Since ENaC is an electrogenic channel its function is highly dependent to cell potential, although a large cystein-containing domain outside the cell indicates a receptorlike purpose[58].

HCO$_3^-$/Cl$^-$

The HCO$_3^-$/Cl$^-$ transporter is gradient driven and exchanges an intracellular bicarbonate ion for an extracellular chloride ion. The electroneutral exchange becomes pH-neutral when coupled to the function of NHE, thus taking up NaCl in exchange for carbondioxide (and water).

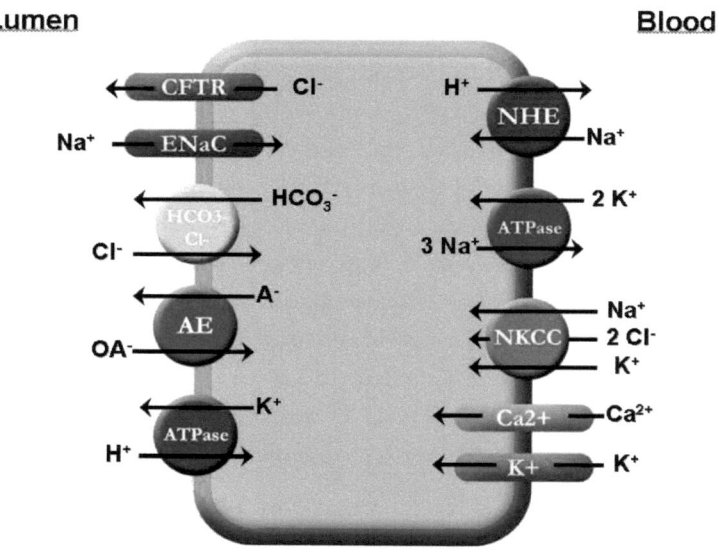

Figure 2: The most important transporters in colonic crypt cells. (Hauser & Schenck 2008)

RVI

Maintaining a constant volume is vital to cells. Perturbations in electrolyte and/or fluid exchange, hormonal influence or other circumstances can cause a change in cell volume. Counteracting swelling of the cell has been collectively termed "regulatory volume decrease" (RVD) and is accomplished by the efflux of potassium-chloride and the resulting water drag[59]. The opposite reaction, the increase of cell volume in response to shrinkage, is termed "regulatory volume increase" (RVI).

Regulatory volume increase is vital to cells that shift water between compartments (i.e. pneumocytes, kidney cells, enterocytes, sweat glands...). In a hypertonic solution, water efflux is counteracted by sodium uptake and its accompanying osmotically driven influx of water. This influx is found to be mainly mediated by sodium-hydrogen-exchangers (NHE) and sodium-potassium-chloride-transporters (NKCC)[60,61]. The importance of NHE1 as the major driver of RVI clearly became visible when the group of Grinstein and Kapus specifically blocked NHE1, and the resulting mutant cells failed to regain volume after exposure to a hypertonic solution[62,63] and even more evident when Rotin and Grinstein examined NHE1 deficient hamster ovary cells, which failed to show significant cell swelling after an osmotic shock induced shrinkage[64]. The anion exchanger (AE) has also been discussed to play a part in RVI but failed to increase volume in NHE1 deactivated sodium-free surrounding[65].

AMPK

The AMP-activated Kinase (AMPK), a serine/threonine protein kinase, is present in all mammalian tissues and has been thoroughly studied in the heart, the skeletal muscle, fat cells, neuronal tissue and the stomach. It is known to be allosterically activated by a high AMP:ATP ratio, hence acting as an energy sensor and regulator of the cell, able to deactivate energy consuming processes and to activate ATP-generating processes. Recent evidence shows that AMPK can also be activated by a calcium-dependent signalling pathway[66]. The artificial agent 5-Aminoimidazole-4-carboxamide-1-beta-4-ribofuranoside (AICAR) can be used to activate AMPK, whereas Compound C acts as an AMPK inhibitor [67,68,69,70,71,72,73,74,75,76,77,78,79,80,81].

This present study demonstrates the presence of AMPK in colonic crypt cells and describes its function as a volume sensor as well as its ability to decrease CFTR dependent chloride efflux and to increase NHE1 dependent sodium influx.

V. HYPOTHESIS

With regards to the literature, 3 hypotheses can be generated:

1) AMPK acts as a sensor for volume and therefore modulates NHE1 activity during RVI.
2) AMPK acts as an inhibitor of CFTR.
3) RVI mediated by NHE1 is independent of CFTR activity.

VI. MATERIAL AND METHODS

Laboratory Animals

Male Sprague-Dawley rats weighing 250-400g (Charles River Laboratories, Wilmington, MA) were housed in climate- and humidity-controlled, light-cycled rooms, fed standard chow with free access to water, and handled according to the humane practices of animal care established by the Yale Animal Care and Use Committee.

Solutions and substances

The composition of the solutions is presented in Figure 3. All data are presented as mmol/l. The osmolarity of all solutions was adjusted to 295 ± 2 mOsm and only changed to 345 ± 2 mOsm for the osmotic shock experiments by adding sucrose. The pH was adjusted to 7.4 for all solutions except the High K^+ Calibration solution, which was adjusted to pH 7.0. Titration of pH was done at 37°C by adding NaOH or HCl. HCl was not used for adjustment of the chloride-free solution and NaOH not for the sodium-free solution. None of the solutions contained bicarbonate. If not otherwise stated, all chemicals were obtained from J.T.Baker, Phillipsburg, NJ, USA.

	Standard HEPES	Sodium-free HEPES	Chloride-free HEPES	High K^+ Calibration
NaCl	115			
HEPES	32.2	32.2	32.2	32.2
KCl	5.0	5		105
MGSO$_4$	1.2	1.2	1.2	1.2
CaCl$_2$	1	1		1
Glucose	10	10	10	
NMDG		132.8		32.8
Mannitol				5
Na$^+$-Gluconate			120	
K$^+$-Gluconate			5	
Ca^{2+}-Gluconate			2	

Figure 3: Composition of solutions used for perfusion of isolated colonic crypts. Data are presented in mmol/l.

Fluorophores

BCECF AM	Molecular Probes, Eugene, OR, USA
Fluo-4 AM	Molecular Probes, Eugene, OR, USA
MQAE	Molecular Probes, Eugene, OR, USA
AICAR	SIGMA-ALDRICH, St. Louis, MO, USA
Amiloride	SIGMA-ALDRICH, St. Louis, MO, USA
Cell-Tak™	Collaborative Research, Bedford, MA, USA
Compound C	SIGMA-ALDRICH, St. Louis, MO, USA
DMSO	J.T.Baker, Phillipsburg, NJ, USA
EDTA	J.T.Baker, Phillipsburg, NJ, USA
Forskolin	SIGMA-ALDRICH, St. Louis, MO, USA
HEPES	SIGMA-ALDRICH, St. Louis, MO, USA
Ionomyecin	Calbiochem, Gibbstown, NJ, USA
Nigericin	Calbiochem, Gibbstown, NJ, USA
NMDG	SIGMA-ALDRICH, St. Louis, MO, USA
NPPB	SIGMA-ALDRICH, St. Louis, MO, USA

Technical equipment

Osmette II osmometer	Precision Systems, Natick, MA, USA
Pinnacle pH meter 530	Corning, Kennebunk, ME, USA
PM400 precision balance	Mettler-Toledo, Columbus, OH, USA
PG2002-S precicion balance	Mettler-Toledo, Columbus, OH, USA
Olympus microscope IX50, IX70	Olympus, Center Valley, PA, USA
C4742-98 CCD-camera	Hamamatsu, Bridgewater, NJ, USA
Lamda DG-4 wavelength switcher	Sutter Instrument, Novato, CA, USA

Software

MetaFluor, V.5.0	Universal Imaging Corp., Downington, PA, USA
Microsoft Excel 2002	Microsoft Corp., Redmond, WA, USA

Methods

Isolation and preparation of colonic crypts

Animals were killed with an overdose of isoflurane anaesthesia and cervical dislocation. An U-shaped abdominal incision was made. The bottom of the U was located approximately 1 cm above the genital region and the upper end at the height of the diaphragm. After dissecting three to four centimeters of descending colon, the colon was longitudinal opened and washed with deionised water to remove residual stool particles. Isolation of single crypts was achieved in adaptation of previously described methods[82,83,84,85]. Therefore the colon was placed in 3 ml of 21 mM EDTA solution at 37°C for three to four minutes and manually shaken for one more minute. 1 ml of this solution was brought to a petri dish containing 40 ml of ice-cold HEPES-buffered Ringer solution, to slow down the activity of the EDTA. Under the microscope, the individual isolated colonic crypts were transferred from the dish to coverslips using a pipette that had been developed for that purpose by Prof. J.P. Geibel M.D., D.Sc.. Prior to that, the coverslips were precoated with Cell-Tak™, a biological adhesive extracted from the marine mussel Mytilus edulis, in order to prevent the crypts to be moved by the perfusion solutions. 0.35 µl of Cell-Tak™ were used to cover an area of approximately one square centimeter of the coverslip. The coverslips were prepared between 20 and 120 minutes before their use and until then stored in a 4°C refrigerator. Several minutes prior to bringing the crypts onto the coverslips, one drop of 4°C HEPES solution was pipetted onto the Cell-Tak™ to activate the glue.

The coverslips with the isolated crypts were mounted on the stage of inverted microscopes (Olympus IX50 and IX70) in a thermostatically controlled acrylic perfusion chamber designed by Prof. J.P. Geibel M.D., D.Sc. The perfusion chamber and an attached perfusion system kept the crypts on the coverslip constantly hydrated by a laminar fluid flow and allowed to change quickly between different solutions. The chamber and all experimental solutions were maintained at 37 ± 0.5 °C during all experiments.

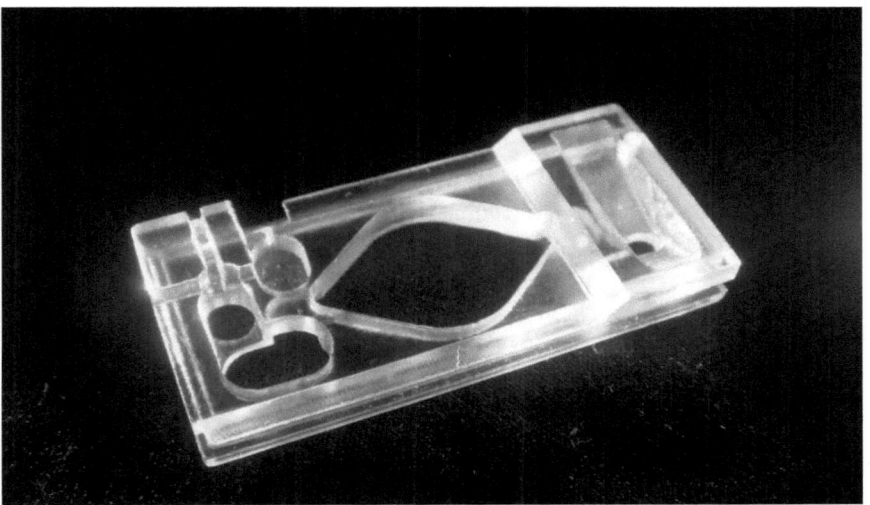

Figure 4: Perfusion Chamber for imaging of vital cells by fluorescence (Schenck & Hauser 2008)

Digital imaging for measurement of intracellular pH, chloride and calcium concentration

Fluorescence describes the optical phenomenon of compounds called fluorophores to absorb a photon and instantly emit a photon of lower energy. The absorbed photon lifts an electron of the compound to a higher but unstable state of energy. Because of the instability of this state the electron returns to the previous lower but stable energy state in less than one millionth of a second. The energy between the two states is released through creating a new photon. Due to the law of conservation of energy, the energy of the emitted photon can not have a higher energy as the absorbed photon. As some energy is lost in the fluorescence process, mainly in heat, the emited photon has a lower energy and therefore higher wavelength than the absorbed photon. The difference of their wavelengths is described by the Stokes Shift.

There are many different fluorophores being industrially produced for a wide range of applications, e.g. fluorophores that help to track compounds by attaching to them, or fluorophores which change their emission rate in dependence to their surrounding. Depending on the molecular structure of the specialized fluorophores, surrounding factors which change their emission rate can be e.g. ionic concentrations, membrane potential, or intracellular pH.

Three different fluorophores were chosen for the experiments described. BCECF AM was used for measuring of intracellular pH, Fluo-4 AM for measuring changes of intracellular calcium concentration and MQAE for measuring changes of intracellular chloride concentration.

BCECF

BCECF (2',7'-bis-(2-carboxyethyl)-5-(and-6)-carboxy-fluorescin-acetomethylester) a dual excitation ratiometric fluorescent pH sensor, allows in vivo intracellular fluorescence imaging of pH[86,87,88]. The fluorescence character of BCECF has helpful differences depending on the excitational wavelength. The excitation of BCECF is pH-independent at lower wavelengths (440nm), whereas at higher wavelengths (490 nm) its excitation is pH-sensitive. Ratios can be calculated from the fluorescence intensities resulting from the two different wavelengths,. These ratios allow to calculate pH_i values that are less susceptible to interferences through bleaching, leakage or uneven loading of the cells[89].

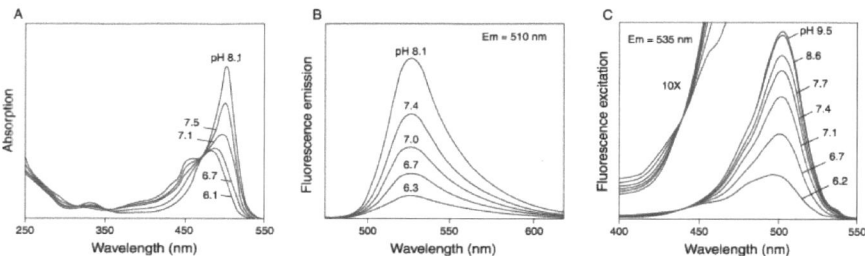

Figure 5: Absorption (A), Emission (B) and Excitation spectra (C) of BCECF[90]

The AM (acetoxymethyl) ester groups of BCECF AM balance the negative charges of BCECF and in this way allow the molecule to cross cell membranes. Intracellular esterases cleave the AM ester groups from BCECF AM. The resulting negatively charged BCECF is now trapped inside the cell, since its diffusion through the cell membrane is inhibited by the also negatively charged membrane. Additionally, the AM ester groups of BCECF AM help to validate cell viability because BCECF AM is not fluorescent until it is converted to BCECF[91]. A 10 mM BCECF AM stock solution was prepared using DMSO and cell loading was performed by incubating the crypts with 1 ml of HEPES buffered 10 μM BCECF-AM for 15 minutes[92,93].

The BCECF emission intensity ratio data (490 nm / 440 nm) was converted to pH_i values using the high K^+/Nigericin calibration technique[94,95,96].

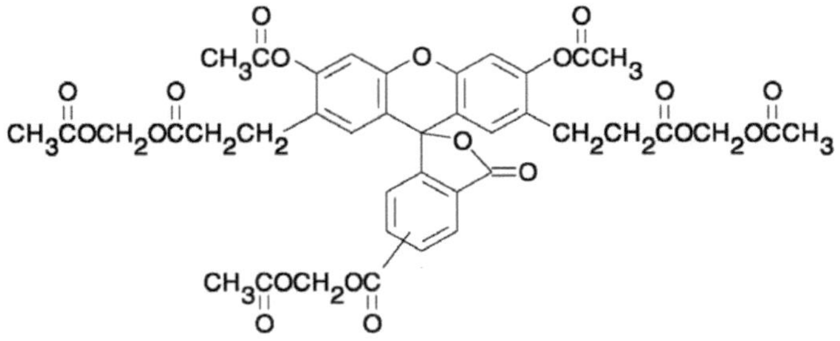

[Chemical structure diagram of BCECF AM]

Figure 6: Molecular structure of one of the three different molecular species of BCECF AM [97]

MQAE

MQAE (N-(ethoxycarbonylmethyl)-6-methoxyquinolinium bromide) is a chloride-sensitive fluorophore[98]. Being a so called quenching dye, its fluorescence intensity is inversely proportional to intracellular chloride concentration without a shift of wavelength[99]. The reciprocal of the concentration of ions that reaches 50% of maximum quenching is called the Stern-Volmer quenching constant (K_{SV}) and is used to describe the efficiency of the quenching process. The K_{SV} of MQAE for chloride is 200 M^{-1} [100]. Fluorescence of MQAE is unaffected by pH or bicarbonate and leakage from cells is slow (<20% in 60 min. at 37°C)[101].

Incubation was performed for 30 minutes with 9.78 mg MQAE in 1 ml of HEPES-buffered Ringer solution (30 mM).

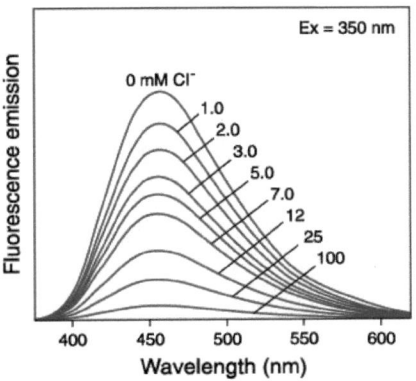

Figure 7: Relation of fluorescence emission and wavelength of MQAE depending on chloride concentration. Increase of chloride concentration[102] leads to decrease of the fluorescence signal.

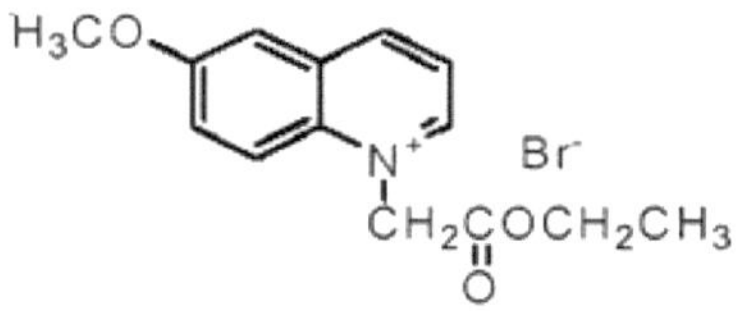

H₃CO — [structure] — N⁺ Br⁻
CH₂COCH₂CH₃
‖
O

Figure 8: Molecular structure of MQAE 103

Fluo-4 AM

The fluorescent dye Fluo-4 AM, the follower of Fluo-3, is designed for detection of intracellular calcium concentrations ($K_d(Ca^{2+})$= 0.35 µM) in the range of 100 nM to 1 µM[104,105].
A 10 mM stock solution of Fluo-4 AM in DMSO was prepared and for incubation 1 ml of 10 µM Fluo-4 AM dissolved in HEPES was used. Crypts were incubated for 15 minutes.
To establish a basis of comparison at the end of Fluo-4 experiments, a perfusion with 2 µM Ionomycin in HEPES was performed to adjust intracellular calcium concentration to extracellular calcium concentration [106,107].

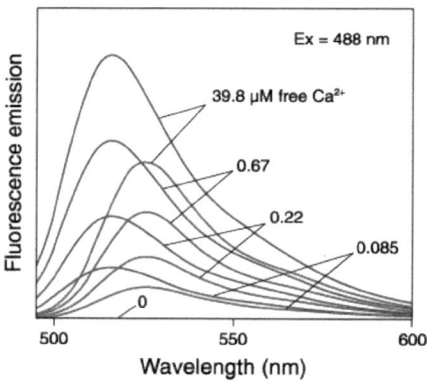

Figure 9: Relation of Fluorescence emission spectra of Fluo-4 (blue) and its precursor Fluo-3 (red) to calcium concentration108

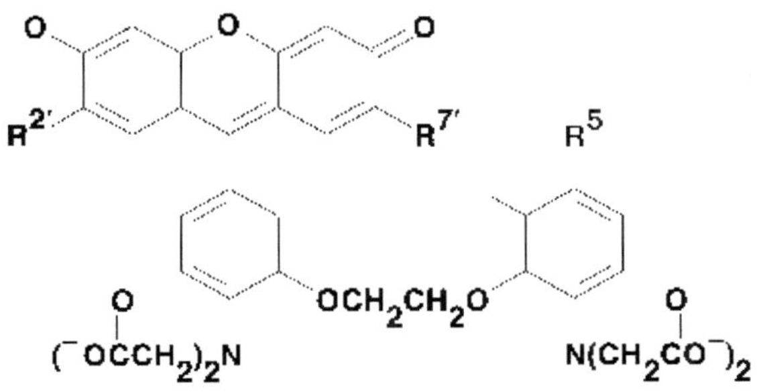

Figure 10: Molecular structure of Fluo Indicators. The side groups of Fluo-4 are: $R2'$ and $R7' = F$; $R5 = CH_3$; $R6 = H$ 109

Incubation

Loading of colonic crypt cells with fluorophores was achieved by allowing the fluorophores to diffuse passively to their intracellular destination. This process was managed at a constant temperature of 37°C in the perfusion chamber, mounted on the stage of the inverted microscopes. 1 ml of the HEPES buffered incubation solution was pipetted into the perfusion chamber. In the absence of light, cells were incubated with the BCECF or Fluo-4 solutions for 15 min, or with the MQAE solution for 30 min.

Following dye-loading, the chamber was flushed for three minutes with a HEPES solution to remove all extracellular dye.

Setup

The fluorescence imaging setup was mounted on a vibration absorbing, light shielded table to minimize noise. The microscopes were used in the epifluorescence mode with 40x objectives. Light from the light source (Lamda DG-4 wavelength switcher) was filtered to select the excitational wavelength specific for the chosen fluorophores. A dichroic mirror was used to direct the excitational light beam to the crypts in the perfusion chamber under the microscope. The resulting fluorescent light of higher wavelength was not reflected by the dichroic mirror but could pass it to be monitored by an intensified charge-coupled device (CCD) camera (Figure 11). Data points were acquired every 15 seconds. Individual regions of interest (columns) were selected by morphology, digitally outlined and monitored during the course of the study. A perfusion system allowed for rapid exhange between different solutions. The perfusion system was temperature controlled and led the solutions directly into the perfusion chamber where a suction device was located at the opposite end for disposal of the solutions. All data, including the individual images for all wavelengths, were recorded to the hard disk for further analysis.

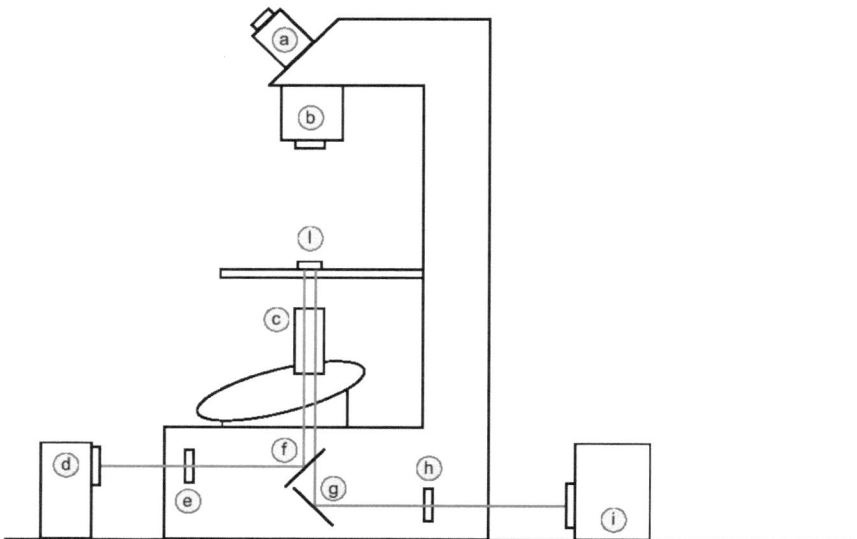

Figure 11: Schematic setup of the inverted fluorescence microscope: a:) ocular, b:) lens, c:) objective, d:) light source, e:) excitational light filter, f:) dichroic mirror, g:) mirror, h:) emission light filter, i:) CCD camera, l:) perfusion chamber with colonic crypts on the coverslip. Perfusion system is not demonstrated. (Schenck & Hauser 2008)

Excitation and monitoring

BCECF was successively excited at 440 nm ± 10 nm and 490 nm ± 10 nm.
The resulting fluorescent signal was monitored at 535 nM ± 10 nm.
MQAE was successively excited at 363 nm ± 10 nm and the resulting fluorescent signal was monitored at 451 nM ± 10 nm.
Fluo-4 AM was successively excited at 490 nm ± 10 nm and the resulting fluorescent signal was monitored at 535 nM ± 10 nm.

Protocols

In six different protocols, the reaction of the cells to changes of their environment was observed by the use of different fluorophores and the high-speed CCD camera. Changes of the cells' environment were managed by switching of the perfusion system. This included changes of osmolarity and changes of electrolyte substitution.

In protocols 1, 2, 4 and 5 crypt cells were exposed to an osmotic shock to cause cell shrinkage. Therefore perfusion was changed to a HEPES solution of higher osmolarity. The osmolarity of this HEPES solution was therefore adjusted from 295 ± 2 to 345 ± 2 mOsm by adding sucrose.

To convert the BCECF emission intensity ratio data (490 nm / 440 nm) to pH_i values, crypts were perfused with the High K^+/Nigericin calibration solution with a pH of 7.0 at the end of every BCECF experiment. The recovery rates are expressed as the ΔpH_i/min and were calculated over the pH_i range of 6 to 8.

Test protocols

Protocol 1

Crypts were superfused with Amiloride containing 295 mOsm HEPES for five minutes. Cells were exposed to an osmotic shock by switching to Amiloride containing 345 mOsm HEPES solution. The resulting trend in pH_i was recorded. Amiloride was used at a concentration of 100 µM. For every observed crypt a control experiment with an identically prepared crypt from the same colon was performed without Amiloride in the solutions. BCECF AM was used for pH_i monitoring.

Protocol 2

Cells were exposed to an osmotic shock and the resulting reaction of pH_i was compared in the presence and absence of 25 µM Compound C (6-[4-(2-piperidin-1-ylethoxy)-phenyl)]-3-pyridin-4-ylpyrazolo[1,5-a] pyrimidine) and 2.5 µL DMSO in the incubation solution. For every observed crypt a control experiment with an identically prepared crypt from the same colon was performed without Compound C in the incubation solution. BCECF AM was used for pH_i monitoring.

Protocol 3

Perfusion of crypts was changed from the HEPES-buffered Ringer solution to a sodium-free solution and switched back to the HEPES-buffered Ringer solution as soon as the pH_i reached equilibrium. This protocol was run in the presence and absence (control) of 25 µM Compound C in the BCECF incubation solution. BCECF AM was used for pH_i monitoring.

Protocol 4

Cells were exposed to an osmotic shock and post osmotic shock pH_i was studied in the presence and absence (control) of the CFTR inhibiting agent NPPB. NPPB was therefore added to the superfusing and the incubation solutions in a concentration of 25 µM. BCECF AM was used for pH_i monitoring.

Protocol 5

Cells were exposed to an osmotic shock and post osmotic shock changes of intracellular calcium were observed by the use of the calcium concentration dependent fluorophore Fluo-4 AM. At the end of the experiment, cells were perfused with 2 µM Ionomycin in HEPES solution to reach an intracellular Ca^{2+} concentration that equals the environment and in this way establishing a standard Ca^{2+} concentration.

Protocol 6

The rate of chloride transport out of the cell was evaluated by using the chloride-sensitive fluorophore MQAE. While monitoring, cells were first superfused with a chloride-free HEPES-buffered solution, followed by 7 to 8 minutes of superfusion with a chloride-containing HEPES-buffered Ringer solution and then again superfusion with the chloride-free solution. The reactions of the cells to the changes of the solutions were observed in the presence and absence (control) of 13.2 mM AICAR (5-Aminoimidazole-4-carboxamide 1-β-D-ribofuranoside) in the MQAE incubation solution. All solutions used in the MQAE protocol contained 10 µM Forskolin.

Statistical analysis

An unpaired, Student's *t*-test with $p < 0.05$ was used to test for significant differences and $p < 0.01$ for very significant differences in pH recovery rates. Data are presented as mean ± SEM.

VII. RESULTS

To explore the cellular mechanisms of volume regulation, colonic crypts were isolated and changes of their intracellular pH, respectively of their intracellular calcium and chloride concentration were observed. An osmotic shock was caused by exposing the cells to a hypertonic extracellular solution and their subsequent regulatory volume increase (RVI) was assessed. The rates of alkalinization and acidification (ΔpH/min) of crypt cells were measured by using high resolution video microscopy to detect intensity changes of the pH-sensitive dye BCECF. The pH_i trend during RVI was described as the (ΔpH_i /min) slope over three minutes, beginning two cycles (30 s) after switching from 295 $\pm$ 2 mOsm to the 344$\pm$ 2 mOsm solution. Post osmotic shock rates of pH_i were measured in the presence of Amiloride, Compound C, NPPB and also after a period of sodium depletion. For every experiment a control experiment with identically treated crypts from the same colon was conducted in the absence of the mentioned drugs.

A calibration curve for BCECF emission intensity ratio data (490 nm / 440 nm) and pH_i values was produced and there from equations for converting were formed (Figure 12).

Converting equations

Setup 1: $2.1564 * (X/X_{(Nigericin)}) + 4.856$

Setup 2: $2.4504 * (X/X_{(Nigericin)}) + 4.5714$

X is the continuously monitored BCECF emission intensity ratio data (490 nm / 440 nm).
$X_{(Nigericin)}$ is the BCECF emission intensity ratio data (490 nm / 440 nm) reached at the end of the experiment after perfusion with the High K^+/Nigericin calibration solution with a pH of 7.0.

Changes of intracellular calcium concentration during the osmotic shock were examined with the fluorophore Fluo-4 AM. Compound C was also used to evaluate the effect of AMP Kinase on CFTR and the resulting chloride trends were monitored with MQAE AM.

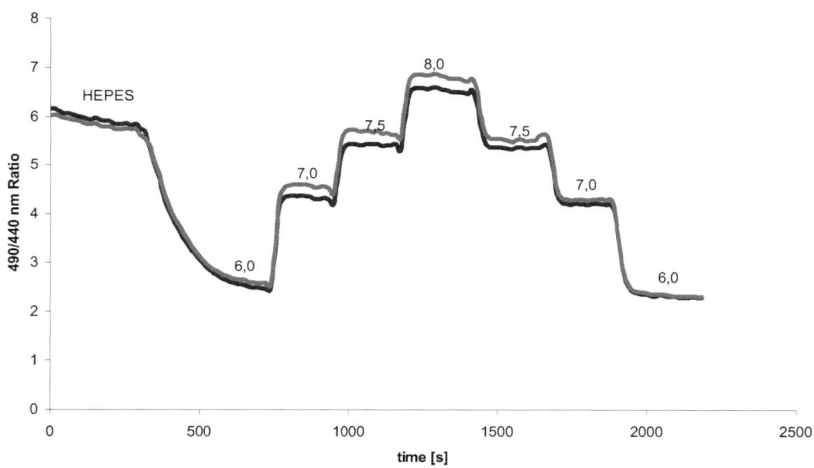

Figure 12: High K+/Nigericin calibration at different levels of pH

Amiloride shows participation of NHE in RVI of colonic crypt cells (Protocol 1)

Osmotic shock causes shrinkage of cells. The cells' physiological response to cell shrinkage is to transport ions into the cell to balance the osmotic gradient and regain their volume by the accompanying drag of water. To what extend NHE plays a role in that reaction to cell shrinkage can be observed experimentally by measuring pH_i. As NHE transports sodium into the cell, it pumps out an equal amount of protons. Hence in the performed experiments NHE activity is proportional to alkalinization of the cell and light emission rate of BCECF. In a physiological environment the pH_i stays stable during NHE activity because of the simultaneous activity of the HCO_3^-/Cl^- cotransporter, which balances the pH_i. In this study, the HCO_3^-/Cl^- cotransporter has been eliminated experimentally by removing HCO_3^- from all buffer solutions. To confirm that the post osmotic shock cell volume recovery and the resulting pH_i change were caused by NHE, the NHE inhibiting agent Amiloride was used. The loss of function of NHE was demonstrated by the cells not being alkalinized after osmotic shock. To have a maximum Amiloride effect at the time of osmotic shock, crypts were superfused with Amiloride containing 295 mOsm HEPES for five minutes before the osmotic shock was performed.

Without Amiloride in the solutions pH$_i$ recovery of 0.0495 ± 0.0072 ΔpH units/min (9 crypts/6 rats) (Figure 13) was observed. Adding Amiloride to the perfusion solutions reduced the rates of alkanization during RVI very significantly (P < 0.01) to 0.0053 ± 0.011 ΔpH units/min (11 crypts/6 rats) (Figure 14).

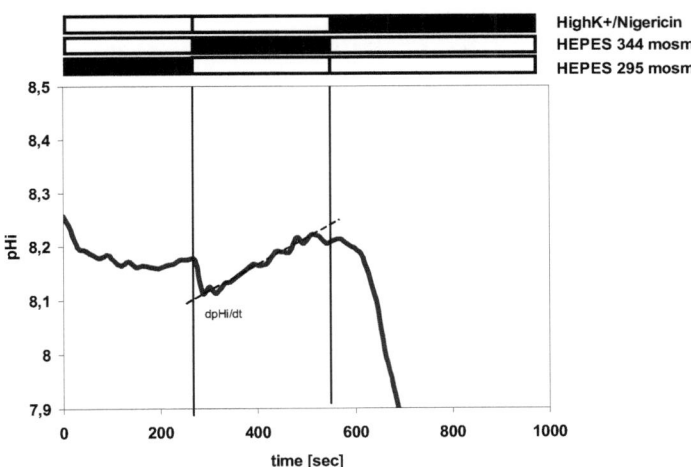

Figure 13: Original trace of a normal volume shift from 295 mosm to 344 mosm in the absence of agents (control). After an initial slight acidification pHi gradually rises due to the activation of NHE. NHE activity was assessed from the slope of the change in pHi within the first three minutes.

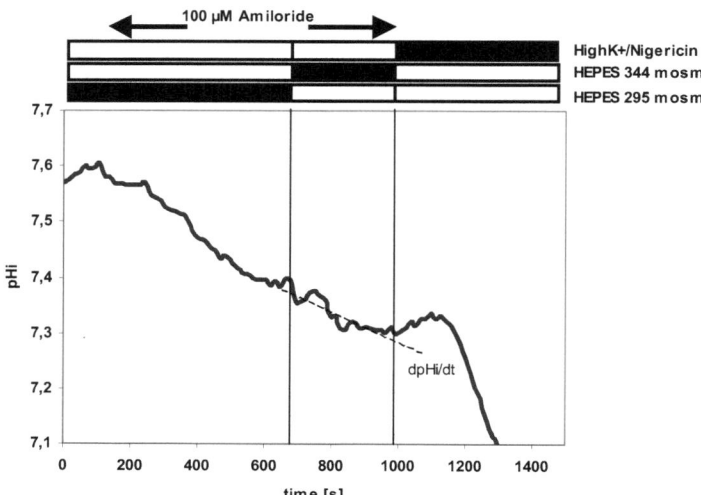

Figure 14: In the presence of 100 µM Amiloride pHi gradually decreases and no recovery can be observed after hypertonic shrinkage.

Compound C suggests effect of AMP Kinase on NHE during RVI (Protocol 2)

AMP Kinase inhibitor Compound C was used to explore how NHE activity during RVI is regulated by AMP Kinase and in this way to find out if AMP Kinase plays a role in the cell's response to shrinkage and thus volume regulation.

The slope of pH_i recovery obtained by BCECF measurements differs very significantly ($P < 0.01$) between 0.040 ± 0.0076 ΔpH units/min (14 crypts/10 rats) in the absence of Compound C and 0.0045 ± 0.0073 ΔpH units/min (14 crypts/10 rats) (Figure 15) in the presence of Compound C in the incubation solution.

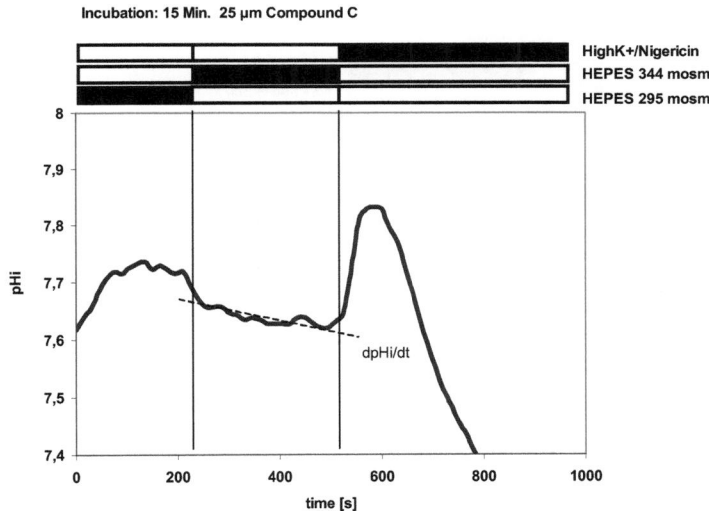

Figure 15: Hypertonicity induced alkanization of the cell is abolished when the crypt is incubated with 25 µM Compound C.

AMP Kinase-dependent NHE regulation is specific for volume changes (Protocol 3)

To evaluate the specifity of AMP Kinase-dependent NHE regulation for volume changes, the activity of NHE as a pH_i regulator in an acidified cells was observed. Acidification of the cells was caused by changing the superfusion from sodium-free to a sodium-containing solution. After switching to a sodium-containing solution the slope of alkalinization was measured over three minutes. The slope did not change significantly ($P = 0.79$) when Compound C was added to the incubation solution to inhibit AMP Kinase-dependent NHE regulation. It varied from 0.2362 ± 0.064 ΔpH units/min (3 crypts/ 3 rats) to 0.2680 ± 0.051 ΔpH units/min (3 crypts/ 3 rats) (Figure 16).

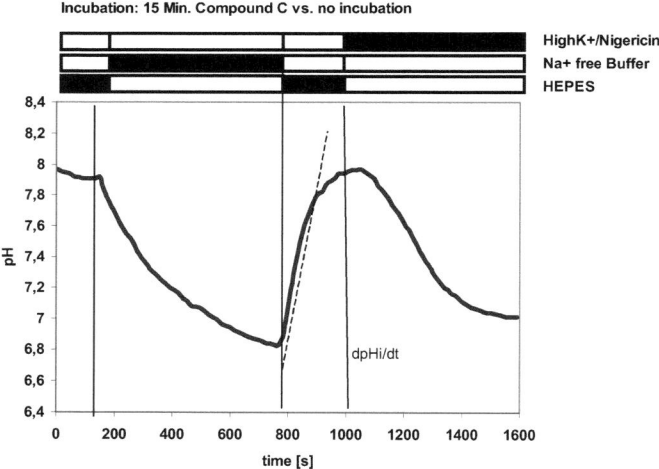

Figure 16: The recovery from acidification of a colonic crypt cell is not influenced by Compound C, indicating that AMPK is a specific activator in volume regulation. The dotted line represents the slope of recovery from acidification, calculated over three minutes after switching to the sodium-containing solution.

Effect of CFTR on RVI (Protocol 4)

In order to see whether CFTR activity affects RVI, pH_i recovery after osmotic shock was compared in the presence and absence of the CFTR inhibiting agent NPPB. The pH_i recovery changed from 0.102 ± 0.089 ΔpH units/min (control: 4 crypts/ 4 rats) to 0.00797 ± 0.0048 ΔpH units/min (NPPB: 4 crypts/ 4 rats) (Figure 17). This change is not significant ($P = 0.35$).

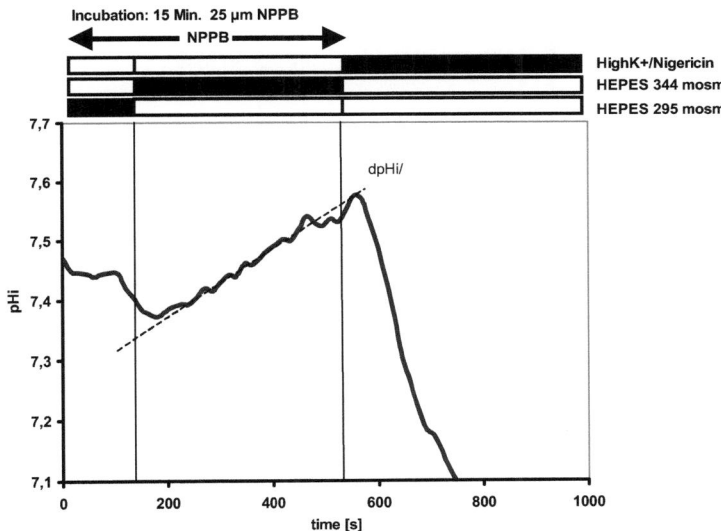

Figure 17: Blocking CFTR by NPPB does not inhibit NHE mediated pHi recovery.

Changes of Calcium concentrations during cell recovery (Protocol 5)

To figure out if RVI is mediated by an increase in intracellular Ca^{2+} concentration the calcium concentration dependent fluorophore Fluo-4 AM was used. In calcium dependent mechanisms, e.g. the depolarization of a skeletal muscle cell, Ca^{2+} concentration rises to the 100 fold of the previous intracellular concentration within a fraction of a second. However, osmotic shock could not trigger a significant amount of Ca^{2+} release. (Figure 18)

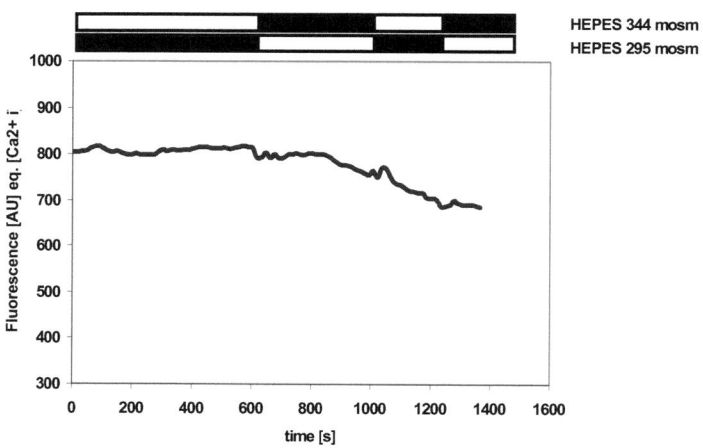

Figure 18: Fluorescence optical measurements of intracellular Ca2+ of Fluo-4 loaded crypt cells (original trace): The dye Fluo-4 was used to detect changes in intracellular Ca2+ after hyperosmotic shock. No consistent changes in Ca2+ concentration could be detected (n = 16).

AMP Kinase and CFTR (Protocol 6)

To detect intracellular concentration changes of chloride, the light emission intensity of MQAE was recorded with high resolution video microscopy. The emission intensity of MQAE is inversely proportional to chloride concentration. The maximum light intensity of the cells was measured by starting with chloride-free solution. As soon as the cells were superfused with the chloride solution, CFTR which was enforced by Forskolin, started to export chloride. The outflow of chloride could only be evaluated when the superfusion was changed to a chloride-free solution, because this eliminated the cell's ability to restore chloride concentration. After switching to chloride-free solution, the slope of fluorescence intensity over time was measured. This slope is inversely proportional to the speed of chloride outflow and allows to evaluate the activity of CFTR.

To evaluate the effect of AMP Kinase on CFTR, the AMP Kinase activator AICAR was added to the incubation solutions. The fluorescence intensity slope, representing the CFTR activity, changed significantly ($P < 0.05$) from 1.04 ± 0.13 without AICAR (10 crypts/10 rats) to 0.78 ± 0.13 in the presence of AICAR (10 crypts/10 rats) (Figure 19).

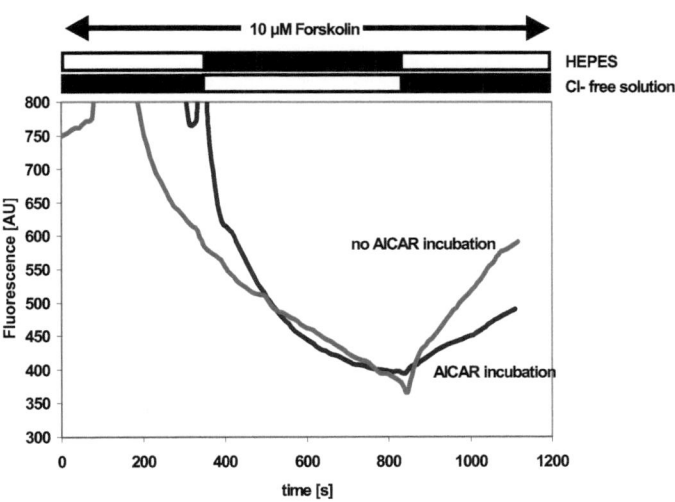

Figure 19: Fluorescence optical measurements of intracellular Cl- of MQAE loaded crypt cells (original trace): Removal of Cl- from the perfusion solutions leads to an increase in the MQAE fluorescence due to cellular exit of Cl- via Forskolin-activated CFTR channels (red trace). After incubation with AICAR, cellular Cl- exit is significantly reduced (blue trace), meaning that AMPK inhibits CFTR. Red trace has been slightly modified in AU and time to fit graph.

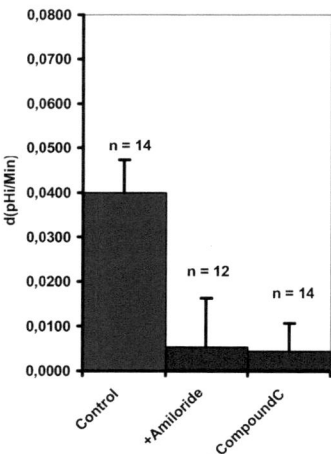

Figure 20: A bar graph of the alkanization rate of the experiments as shown in Fig. 13, 14 and 15. Data are presented as mean ± SEM.

VIII. DISCUSSION

Maintaining a stable cell volume is a crucial task for cells throughout the body. Especially in the colon, where massive fluid transport occurs, cells must be capable to react fast and accurate to volume and osmolarity changes. Many transporters and their integration within signaling cascades have been subject to research in the field of cell volume regulation so far.

NHE1 is one of the transporters known to play an important role in the regulation of cell volume. This effect can be seen experimentally when 100 µM Amiloride is added to block all NHE1 activity which prevents expected alkalinization during RVI (Figures 13 and 14). A similar trend can be seen when the cells were incubated with the specific AMPK inhibitor Compound C before the experiment (Figure 15), showing that AMPK mediates RVI via NHE.

Protocol 3 demonstrates that activity of NHE in an acidified cell is not affected by AMP Kinase, indicating that AMP Kinase controls NHE only in terms of volume regulation. (Figure 16)

AMPK has been proven to inhibit CFTR in a number of tissues (e.g. T84 cell line[110], Xenopus Oocyts[111]). By using the Cl⁻-sensitive dye MQAE in RVI experiments, a decrease in CFTR activity was detected when the AMPK activating agent AICAR was applied. That proves that in colonic crypt cells CFTR is inhibited by AMPK. This finding is in line with the previous results, showing that AMPK triggers Na^+ uptake and inhibits Cl^- secretion to gain cell volume in RVI.

CFTR has been postulated to have multiple regulatory functions in the cell (e.g. Na^+ absorption, HCO_3^--secretion[112,113,114]). CFTR was blocked by the specific inhibitor NPPB in order to assess whether NHE activity in RVI is regulated by CFTR activity. It turned out that alkalinization of the crypt cell in RVI did not change in the presence of NPPB. This indicates that NHE and its activation through AMPK during RVI does not depend on CFTR activity. (Figure 17)

In its function as a second messenger, intracellular Ca^{2+} intake triggers multiple actions in a number of tissues, such as contraction in muscle cells and exocytosis in synapses. To assess whether Ca^{2+} plays a crucial role in RVI of the colonic crypt cell, the Ca^{2+} sensitive dye Fluo-4 AM was used to detect changes in intracellular Ca^{2+} concentration. However, in spite of a high sample number (n = 14) no significant change in Ca^{2+} concentration during RVI was observed. (Figure 18)

IX. CONCLUSIONS

Cell volume regulation is a delicate process that involves multiple pathways to maintain physiologic function of the cell during osmotic distress. In this study a new conductor of this regulation, the AMP-activated Kinase, which regulates RVI by activating NHE and inhibiting CFTR was identified. In this way AMP Kinase activation does not only lead to ion transport into the cell, but also inhibits ion transport out of the cell, thus increasing intracellular osmolarity. In this way AMPK serves as a key regulator of volume regulation in colonic crypt cells, enabling them to perform powerful RVI in hyperosmotic environments. Furthermore, the AMPK action on NHE appears to be specific for volume regulation and does not include changes in intracellular calcium in its pathways. Therefore all three Hypotheses are corroborated.

X. LIST OF ABBREVIATIONS

AE	Anion Exchanger
AICAR	5-Aminoimidazole-4-carboxamide 1-β-D-ribofuranoside
AM	Acetoxymethyl
AU	Arbitrary Units
AMP	Adenosine Monophosphate
AMPK	AMP-activated kinase
ATP	Adenosine Triphosphate
BCECF	2',7'-bis-(2-carboxyethyl)-5-(and-6)-carboxy-fluorescin-acetomethylester
cAMP	Cyclic AMP
CCD	Charge-Coupled Device
cGMP	Cyclic GMP
CFTR	Cystic Fibrosis Transmembrane Conductance Regulator
EDTA	Ethylen-diamin-tetra-acidic acid
ENaC	Epithelial sodium Channel
HEPES	4-(2-hydroxyethyl) piperazine-1-ethanesulfonic acid
mOsm	MilliOsmolarity
MQAE	N-(ethoxycarbonylmethyl)-6-methoxyquinolinium bromide
NHE	Sodium-Hydrogen-Exchanger
NHE1	Sodium-Hydrogen-Exchanger, isoform 1
NMDG	N-methyl-D-glucosamine
NKCC-1	$Na^+,K^+,2Cl^-$ Cotransporter, isoform 1
NPPB	5-nitro-2-(3-phenylpropylamino) benzoic acid
ORT	Oral Rehydration Therapy
PKA	Protein kinase A
PKC	Protein kinase C
RVD	Regulatory volume decrease
RVI	Regulatory volume increase
SEM	Standard of means
WHO	World Health Organization

XI. REFERENCES

Figures

Figure 1: A colonic crypt separated by the EDTA method and colored by BCECF in 400x magnification (Hauser & Schenck 2007). 5
Figure 2: The most important transporters in colonic crypt cells. (Hauser & Schenck 2008) 8
Figure 3: Composition of solutions used for perfusion of isolated colonic crypts. Data are presented in mmol/l. 10
Figure 4: Perfusion Chamber for imaging of vital cells by fluorescence (Schenck & Hauser 2008) 13
Figure 5: Absorption (A), Emission (B) and Excitation spectra (C) of BCECF 14
Figure 6: Molecular structure of one of the three different molecular species of BCECF AM 15
Figure 7: Relation of fluorescence emission and wavelength of MQAE depending on chloride concentration. Increase of chloride concentration leads to decrease of the fluorescence signal. 15
Figure 8: Molecular structure of MQAE 16
Figure 9: Relation of Fluorescence emission spectra of Fluo-4 (blue) and its precursor Fluo-3 (red) to calcium concentration. 16
Figure 10: Molecular structure of Fluo Indicators. The sidegroups of Fluo-4 are: R2' and R7'= F; R5 = CH_3; R6 = H 17
Figure 11: Schematic setup of the inverted fluorescence microscope: a:) ocular, b:) lens, c:) objective, d:) lightsource, e:) excitational light filter, f:) dichroic mirror, g:) mirror, h:) emission light filter, i:) CCD camera, l:) perfusion chamber with colonic crypts on the coverslip. Perfusion system is not demonstrated. (Schenck & Hauser 2008) 18
Figure 12: High K^+/Nigericin calibration at different levels of pH 22
Figure 13: Original trace of a normal volumeshift from 295 mosm to 344 mosm in the absence of agents (control). After an initial slight acidification pH_i gradually rises due to the activation of NHE. NHE activity was assessed from the slope of the change in pH_i within the first three minutes. 23
Figure 14: In the presence of 100 μM Amiloride pH_i gradually decreases and no recovery can be observed after hypertonic shrinkage. 24
Figure 15: Hypertonicity induced alkanization of the cell is abolished when the crypt is incubated with 25 μM Compound C 25

Figure 16: The recovery from acidification of a colonic crypt cell is not influenced by Compound C, indicating that AMPK is a specific activator in volume regulation. The dotted line represents the slope of recovery from acidification, calculated over three minutes after switching to the sodium-containing solution. 26

Figure 17: Blocking CFTR by NPPB does not inhibit NHE mediated pH_i recovery 27

Figure 18: Fluorescence optical measurements of intracellular Ca^{2+} of Fluo-4 loaded crypt cells (original trace): The dye Fluo-4 was used to detect changes in intracellular Ca^{2+} after hyperosmotic shock. No consistent changes in Ca^{2+} concentration could be detected (n = 16). 28

Figure 19: Fluorescence optical measurements of intracellular Cl⁻ of MQAE loaded crypt cells (original trace): Removal of Cl⁻ from the perfusion solutions leads to an increase in the MQAE fluorescence due to cellular exit of Cl⁻ via Forskolin-activated CFTR channels (red trace). After incubation with AICAR, cellular Cl⁻ exit is significantly reduced (blue trace), meaning that AMPK inhibits CFTR. Red trace has been slightly modified in AU and time to fit graph. 29

Figure 20: A bar graph of the alkanization rate of the experiments as shown in Fig. 13, 14 and 15. Data are presented as mean ± SEM. 30

Literature

1 Drake RL, Vogl W, Mitchell AWM: Gray's anatomy for students. Philadelphia, Elsevier, 2005; p279-314.

2 Ross MH, Pawlind W: Histology: A text and atlas, Fifth edition, Philadelphia, Lippincott & Williams. 2006.

3 Junqueira, Carneiro, Kelley: Histologie, 5.te Auflage, München, Springer 2002; p263-266.

4 Debongnie JC and Phillips SF: Capacity of the human colon to absorb fluid. Gastroenterology 1978; 74: 698-703.

5 Warth R and Bleich M: K+ channels and colonic function. Rev Physiol Biochem Pharmacol. 2000; 140: 1-62.

6 Geibel JP, Rajendran VM and Binder HJ: Na(+) dependent fluid absorption in intact perfused rat colonic crypts, gastroenterology 2001; 120:144-150.

7 Greger R: Role of CFTR in the colon, Review, Annu Rev. Physiol. 2000. 62:467-91.

8 Grinstein S, Clarke CA, Rothstein A: Activation of Na+/H+ exchange in lymphocytes by osmotically induced volume changes and by cytoplasmic acidification. J Gen Physiol. Nov1983; 82(5):619-38.

9   Demaurex N and Grinstein S: Na+/H+ antiport: modulation by ATP and role in cell volume regulation. J Exp Biol. 1994 Nov; 196:389-404.

10 Orlowski J, Kandasamy RA: Delineation of transmembrane domains of the Na+/H+ exchanger that confer sensitivity to pharmacological antagonists. J Biol Chem. 1996; Aug 16;271(33):19922-7.

11 Shrode LD, Gan BS, D'Souza SJ, Orlowski J, Grinstein S: Topological analysis of NHE1, the ubiquitous Na+/H+ exchanger using chymotryptic cleavage. Am J Physiol. 1998; Aug;275(2 Pt 1):C431-9.

12 Tse CM, Ma AI, Yang VW, Watson AJ, Levine S, Montrose MH, Potter J, Sardet C, Pouyssegur J, Donowitz M: Molecular cloning and expression of a cDNA encoding the rabbit ileal villus cell basolateral membrane Na+/H+ exchanger. EMBO J. 1991; 10:1957–67.

13 Orlowski J, Kandasamy RA, Shull GE: Molecular cloning of putative members of the Na/H exchanger gene family. cDNA cloning, deduced amino acid sequence, and mRNA tissue expression of the rat Na/H exchanger NHE-1 and two structurally related proteins. J. Biol. Chem. 1992; 267:9331–39.

14 Tse CM, Levine SA, Yun CH, Montrose MH, Little PJ, Pouyssegur J, Donowitz M: Cloning and expression of a rabbit cDNA encoding a serum-activated ethylisopropylamilorideresistant epithelial Na+/H+ exchanger isoform (NHE2). J. Biol. Chem. 1993; 268:11917–24.

15 Tse CM, Brant SR, Walker MS, Pouyssegur J, Donowitz M: Cloning and sequencing of a rabbit cDNA encoding an intestinal and kidney-specific Na+/H+ exchanger isoform (NHE-3). J. Biol. Chem. 1992; 267:9340–46.

16 Klanke CA, Su YR, Callen DF, Wang Z, Meneton P, Baird N, Kandasamy RA, Orlowski J, Otterud BE, Leppert M: Molecular cloning and physical and genetic mapping of a novel human Na+/H+ exchanger (NHE5/SLC9A5) to chromosome 16q22.1. Genomics 1995; 25:615–22.

17 Baird NR, Orlowski J, Szabo EZ, Zaun HC, Schultheis PJ, Menon AG, Shull GE: Molecular cloning, genomic organization, and functional expression of Na+/H+ exchanger isoform 5 (NHE5) from human brain. J.Biol. Chem. 1999; 274:4377–82.

18 Numata M, Petrecca K, Lake N, Orlowski J: Identification of a mitochondrial Na+/H+ exchanger. J. Biol. Chem. 1998; 273: 6951–59.

19 Brett CL, Wei Y, Donowitz M, Rao R: Human Na+/H+ exchanger isoform 6 is found in ecycling endosomes of cells, not in mitochondria. Am. J. Physiol. Cell Physiol. 2002; 282:C1031–41.

20 Numata M, Orlowski J: Molecular cloning and characterization of a novel (Na+,K+)/H+ exchanger localized to the trans-Golgi network. J. Biol. Chem. 2001; 276:17387–94.

21 Goyal S, Vanden Heuvel G, Aronson PS: Renal expression of novel Na+/H+ exchanger isoform NHE8. Am. J. Physiol.Renal Physiol. 2003; 284:F467–73.

22 de Silva MG, Elliott K, Dahl HH, Fitzpatrick E, Wilcox S, et al.: Disruption of a novel member of a sodium/hydrogen exchanger family and DOCK3 is associated with an attention deficit hyperactivity disorder-like phenotype. J. Med. Genet. 2003; 40:733–40.

23 Cala PM, Maldonado HM: pH regulatory Na/H exchange by Amphiuma red blood cells. J Gen Physiol. 1994 Jun;103(6):1035-53.

24 Parker JC, McManus TJ, Starke LC, Gitelman HJ: Coordinated regulation of Na/H exchange and [K-Cl] cotransport in dog red cells. J Gen Physiol. 1990 Dec; 96(6):1141-52.

25 Parker JC, Dunham PB, Minton AP: Effects of ionic strength on the regulation of Na/H exchange and K-Cl cotransport in dog red blood cells. J Gen Physiol. 1995 Jun; 105(6):677-99.

26 Ericson AC, Spring KR: Volume regulation by Necturus gallbladder: apical Na+-H+ and Cl(-)- HCO-3 exchange. Am J Physiol. 1982 Sep; 243(3):C146-50.

27 Cheung RK, Grinstein S, Dosch HM, Gelfand EW: Volume regulation by human lymphocytes: characterization of the ionic basis for regulatory volume decrease. J Cell Physiol. 1982 Aug; 112(2):189-96.

28 Grinstein S, Cohen S, Goetz JD, Rothstein A: Na+/H+ exchange in volume regulation and cytoplasmic pH homeostasis in lymphocytes. Fed Proc. 1985 Jun; 44(9):2508-12.

29 Alexander RT and Grinstein S: Na+/H+ exchangers and the regulation of volume, review, Acta Physiol 2006; 187, p159-167.

30 Garnovskaya MN, Mukhin YV, Vlasova TM, Raymond JR: Hypertonicity activates Na+/H+ exchange through Janus kinase 2 and calmodulin. J Biol Chem. 2003 May 9; 278(19):16908-15.

31 Grinstein S, Woodside M, Sardet C, Pouyssegur J, Rotin D: Activation of the Na+/H+ antiporter during cell volume regulation. Evidence for a phosphorylation-independent mechanism. J Biol Chem. 1992 Nov 25; 267(33):23823-8.

32 McSwine RL, Li J, Villereal ML: Examination of the role for Ca2+ in regulation and phosphorylation of the Na+/H+ antiporter NHE1 via mitogen and hypertonic stimulation. J Cell Physiol. 1996 Jul;168(1):8-17.

33 Krump E, Nikitas K, Grinstein S: Induction of tyrosine phosphorylation and Na+/H+ exchanger activation during shrinkage of human neutrophils. J Biol Chem. 1997 Jul 11; 272(28):17303-11.

34. Gillis D, Shrode LD, Krump E, Howard CM, Rubie EA, Tibbles LA, Woodgett J, Grinstein S: Osmotic stimulation of the Na+/H+ exchanger NHE1: relationship to the activation of three MAPK pathways. J Membr Biol. 2001 Jun 1; 181(3):205-14.

35. Krump E, Nikitas K, Grinstein S: Induction of tyrosine phosphorylation and Na+/H+ exchanger activation during shrinkage of human neutrophils. J Biol Chem. 1997 Jul 11; 272(28):17303-11.

36. Fuster D, Moe OW, Hilgemann DW: Lipid- and mechanosensitivities of sodium/hydrogen exchangers analyzed by electrical methods. Proc Natl Acad Sci U S A. 2004 Jul 13; 101(28):10482-7.

37. Ikuma M, Kashgarian M, Binder HJ, Rajendran VM: Differential regulation of NHE isoforms by sodium depletion in proximal and distal segments of rat colon. Am J Physiol. 1999 Feb; 276(2 Pt 1):G539-49.

38. Hasselblatt P, Warth R, Schulz-Baldes A, Greger R, Bleich M: pH regulation in isolated in vitro perfused rat colonic crypts. Pflugers Arch. 2000 Nov; 441(1):118-24.

39. Bachmann O, Riederer B, Rossmann H, Groos S, Schultheis PJ, Shull GE, Gregor M, Manns MP, Seidler U: The Na+/H+ exchanger isoform 2 is the predominant NHE isoform in murine colonic crypts and its lack causes NHE3 upregulation. Am J Physiol Gastrointest Liver Physiol. 2004 Jul; 287(1):G125-33.

40. Biemesderfer D, Pizzonia J, Abu-Alfa A, Exner M, Reilly R, Igarashi P, Aronson PS: NHE3: a Na+/H+ exchanger isoform of renal brush border. Am J Physiol. 1993 Nov; 265(5 Pt 2):F736-42.

41. Biemesderfer D, Rutherford PA, Nagy T, Pizzonia JH, Abu-Alfa AK, Aronson PS: Monoclonal antibodies for high-resolution localization of NHE3 in adult and neonatal rat kidney Am J Physiol. 1997 Aug; 273(2 Pt 2):F289-99.

42. Hoogerwerf WA, Tsao SC, Devuyst O, Levine SA, Yun CH, Yip JW, Cohen ME, Wilson PD, Lazenby AJ, Tse CM, Donowitz M: NHE2 and NHE3 are human and rabbit intestinal brush-border proteins. Am J Physiol. 1996 Jan; 270(1 Pt 1):G29-41.

43. Noel J, Roux D, Pouysségur J: Differential localization of Na+/H+ exchanger isoforms (NHE1 & NHE3) in polarized epithelial cell lines J Cell Sci. 1996 May; 109 (Pt 5):929-39.

44. Schultheis PJ, Clarke LL, Meneton P, Harline M, Boivin GP, Stemmermann G, Duffy JJ, Doetschman T, Miller ML, Shull GE: Targeted disruption of the murine Na+/H+ exchanger isoform 2 gene causes reduced viability of gastric parietal cells and loss of net acid secretion. J Clin Invest. 1998 Mar 15; 101(6):1243-53.

45 Drumm K, Lee E, Stanners S, Gassner B, Gekle M, Poronnik P, Pollock C: Albumin and glucose effects on cell growth parameters, albumin uptake and Na(+)/H(+)-exchanger Isoform 3 in OK cells. Cell Physiol Biochem. 2003; 13(4):199-206.

46 Miyazaki E, Sakaguchi M, Wakabayashi S, Shigekawa M, Mihara K: NHE6 protein possesses a signal peptide destined for endoplasmic reticulum membrane and localizes in secretory organelles of the cell. J Biol Chem. 2001 Dec 28; 276(52):49221-7.

47 Brett CL, Wei Y, Donowitz M, Rao R: Human Na(+)/H(+) exchanger isoform 6 is found in recycling endosomes of cells, not in mitochondria. Am J Physiol Cell Physiol. 2002 May; 282(5):C1031-41.

48 Nakamura N, Tanaka S, Teko Y, Mitsui K, Kanazawa H: Four Na+/H+ exchanger isoforms are distributed to Golgi and post-Golgi compartments and are involved in organelle pH regulation. J Biol Chem. 2005 Jan 14; 280(2):1561-72.

49 Riordan JR, Rommens JM, Kerem B-S, Alon N, Rozmahel R, Grzelczak Z, Zielenski J, Plavsic SLN, Chou J-L, Drumm ML, Iannuzzi CM, Collins FS, Tsui L-C: Identification of the cystic fibrosis gene: cloning and characterization of complementary DNA. Science 1989; 245: 1066–1072.

50 Kunzelmann K and Hall M: Electrolyte Transport in the Mammalian Colon: Mechanisms and Implications for Disease. Physiol Rev. 2002; 82: 245–289.

51 Hallows KR, Kobinger GP, Wilson JM, Witters LA and Foskett JK: Physiological modulation of CFTR activity by AMP-activated protein kinase in polarized T84 cells. Am J Physiol Cell Physiol. 2003; 284:1297-1308.

52 Rajendran VM, Geibel JP, Binder HJ: Role of Cl channels in Cl-dependent Na/H exchange, Am J Physiol. 1999 Jan; 276(1 Pt 1):G73-8.

53 Rajendran VM, Geibel JP, Binder HJ: chloride-dependent Na-H exchange. A novel mechanism of sodium transport in colonic crypts. J Biol Chem. 1995; 12;270(19):11051-4.

54 Greger R: Role of CFTR in the colon. Annu Rev Physiol. 2000; 62:467-91.

55 Welsh MJ, Tsui LC, Boat TF, Beudet AL: Cystic Fibrosis. In The Metabolic and Molecular Basis of Inherited Disease. New York: McGraw-Hill Inc. 1995; 3799-3876.

56 Dunhan PB, Jessen F, Hoffmann EK: Inhibition of Na-K-Cl cotransport in Ehrlich ascites cells by antiserum against purified proteins of the cotransporter. Proc. Natl. Acad. Sci. USA. 1990; 87(17): 6828–6832.

57 Geck P, Pfeiffer B: Na/K/2Cl -cotransport in animal cells: its role in volume regulation. Ann. N Y Acad. Sci. 1985; 456: 166–182.

58 Kunzelmann K and Hall M: Electrolyte Transport in the Mammalian Colon: Mechanisms and Implications for Disease. Physiol Rev. 2002; 82: 245–289.

59 Graf J, Haddad P, Haeussinger D, Lang F: Cell volume regulation in liver. Ren Physiol Biochem. 1988 May-Oct; 11(3-5):202-20.
60 Lang F, Busch GL, Völkl H: The diversity of volume regulatory mechanisms. Cell Physiol Biochem. 1998; 8(1-2):1-45.
61 Waldegger S, Steuer S, Risler T, Heidland A, Capasso G, Massry S, Lang F: Mechanisms and clinical significance of cell volume regulation. Nephrol Dial Transplant. 1998 Apr; 13(4):867-74.
62 Grinstein S, Woodside M, Sardet C, Pouyssegur J, Rotin D: Activation of the Na+/H+ antiporter during cell volume regulation. Evidence for a phosphorylation-independent mechanism. J Biol Chem. 1992 Nov 25; 267(33):23823-8.
63 Kapus A, Grinstein S, Wasan S, Kandasamy R, Orlowski J: Functional characterization of three isoforms of the Na+/H+ exchanger stably expressed in Chinese hamster ovary cells. ATP dependence, osmotic sensitivity, and role in cell proliferation. J Biol Chem. 1994 Sep 23; 269(38):23544-52.
64 Rotin D, Grinstein S: Impaired cell volume regulation in Na(+)-H+ exchange-deficient mutants. Am J Physiol. 1989 Dec; 257(6 Pt 1):C1158-65.
65 Humphreys BD, Jiang L, Chernova MN, Alper SL: Hypertonic activation of AE2 anion exchanger in Xenopus oocytes via NHE-mediated intracellular alkalinization. Am J Physiol. 1995 Jan; 268(1 Pt 1):C201-9.
66 Birnbaum MJ: Activating AMP-Activated Protein Kinase without AMP; Preview; Molecular Cell 2005 August 5; Vol. 19, 289–296.
67 Carling D: The AMP-activated protein kinase cascade--a unifying system for energy control. Trends Biochem Sci. 2004; 29:18-24.
68 Corton JM, Gillespie JG, Hawley SA, Hardie DG: 5-aminoimidazole-4-carboxamide ribonucleoside. A specific method for activating AMP-activated protein kinase in intact cells? Eur J Biochem. 1995; 229:558-565.
69 Davies SP, Carling D, Hardie DG: Tissue distribution of the AMP-activated protein kinase, and lack of activation by cyclic-AMP-dependent protein kinase, studied using a specific and sensitive peptide assay. Eur J Biochem. 1989; 186:123-128.
70 Hardie DG, Scott JW, Pan DA, Hudson ER: Management of cellular energy by the AMPactivated protein kinase system. FEBS Lett. 2003; 546:113-120.
71 Hardie DG. Minireview: the AMP-activated protein kinase cascade: the key sensor of cellular energy status. Endocrinology 2003;144:5179-5183.
72 Hawley SA, Davison M, Woods A, Davies SP, Beri RK, Carling D, Hardie DG: Characterization of the AMP-activated protein kinase kinase from rat liver and identification

of threonine 172 as the major site at which it phosphorylates AMP-activated protein kinase. J Biol Chem. 1996; 271:27879-27887.

73 Kahn BB, Alquier T, Carling D, Hardie DG: AMP-activated protein kinase: ancient energy gauge provides clues to modern understanding of metabolism. Cell Metab. 2005; 1:15-25.

74 Kemp BE, Stapleton D, Campbell DJ, Chen ZP, Murthy S, Walter M, Gupta A, Adams JJ, Katsis F, van DB, Jennings IG, Iseli T, Michell BJ, Witters LA: AMP-activated protein kinase, super metabolic regulator. Biochem Soc Trans 2003; 31:162-168.

75 Leff T: AMP-activated protein kinase regulates gene expression by direct phosphorylation of nuclear proteins. Biochem Soc Trans. 2003; 31:224-227.

76 McCullough LD, Zeng Z, Li H, Landree LE, McFadden J, Ronnett GV: Pharmacological inhibition of AMP-activated protein kinase provides neuroprotection in stroke. J Biol Chem. 2005; 280:20493-20502.

77 Stein SC, Woods A, Jones NA, Davison MD, Carling D: The regulation of AMP-activated protein kinase by phosphorylation. Biochem J. 2000; 345 Pt 3:437-443.

78 Woods A, Vertommen D, Neumann D, Turk R, Bayliss J, Schlattner U, Wallimann T, Carling D, Rider MH: Identification of phosphorylation sites in AMP-activated protein kinase (AMPK) for upstream AMPK kinases and study of their roles by site-directed mutagenesis. J Biol Chem. 2003; 278:28434-28442.

79 Henin, N., M. F. Vincent, H. E. Gruber, and G. Van den Berghe: Inhibition of fatty acid and cholesterol synthesis by stimulation of AMP-activated protein kinase. FASEB J. 1995; 9: 541-546.

80 Henin, N., M. F. Vincent, and G. Van den Berghe: Stimulation of rat liver AMP-activated protein kinase by AMP analogues. Biochim Biophys Acta. 1996;1290: 197-203.

81 Sullivan, J. E., K. J. Brocklehurst, A. E. Marley, F. Carey, D. Carling, and R. K Beri: Inhibition of lipolysis and lipogenesis in isolated rat adipocytes with AICAR, a cell-permeable activator of AMPactivated protein kinase. FEBS. 1994; Lett. 353: 33-36.

82 Ikuma M, Binder HJ and Geibel J: Role of Apical H-K Exchange and Basolateral K Channel in the Regulation of Intracellular pH in Rat Distal Colon Crypt Cells. J. Membrane Biol. 1998; 166, 205-212.

83 Del Buono R, Lee CY, Hawkey CJ and Wright NA: Isolated crypts form spheres prior to full intestinal differentiation when grown as xenografts: an in vivo model for the study of intestinal differentiation and crypt neogenesis, and for the abnormal crypt architecture of juvenile polyposis coli. J. Pathol. 2005; 206, 395-401.

84 Rajendran MR, Geibel J and Binder HJ: Chloride-dependent Na-H Exchange. J Biol Chem. 1995; 270, No. 19, 11051-11054. 46

85 Weiser MM: Intestinal Epithelial Cell Surface Membrane Glycoprotein Synthesis. J Biol Chem. 1973; 248, No 7, 2536-2541.
86 Rink TJ, Tsien RY and Pozzan T: Cytoplasmic pH and Free Mg2+ in Lymphocytes. J Cell Biol. 1982; 95(1), 189-96.
87 Ozkan P and Mutharasan R: A rapid method for measuring intracellular pH using BCECF-AM. Biochimica et Biophysica Acta (BBA) 2002; Volume 1572, Issue 1, 143-148.
88 Weinlich M, Heydasch U, Mooren F and Starlinger M: Simultaneous detection of cell volume and intracellular pH in isolated rat duodenal cells by confocal microscopy and BCECF. Res Exp Med (Berl.) 1998; 198(2), 73-82.
89 Molecular Probes: BCECF Manual MP 01150. updated: April 24, 2006; date of access: April, 8 2008. http://probes.invitrogen.com/media/pis/mp01150.pdf
90 Molecular Probes: Online Handbook, Figure 20.3 updated: January 22, 2006; date of access: April, 8 2008. http://probes.invitrogen.com/handbook/figures/0573.html
91 Molecular Probes: BCECF Manual MP 01150, 2006 Updated: April 24, 2006; date of access: April, 8 2008. http://probes.invitrogen.com/media/pis/mp01150.pdf
92 Geibel JP, Wagner CA, Caroppo R, Qureshi I, Gloeckner J, Manuelidis L, Kirchhoff P and Radebold K: The Stomach Divalent Ion-sensing Receptor SCAR Is a Modulator of Gastric Secretion. J Biol Chem. 2001; 276, No. 43, 39549-39552.
93 Waisbren SJ, Geibel J, Boron WF and Modlin IM: Luminal perfusion of isolated gastric glands. Am J Physiol. 1994; 266 C1013-27.
94 Geibel JP, Wagner CA, Caroppo R, Qureshi I, Gloeckner J, Manuelidis L, Kirchhoff P and Radebold K: The Stomach Divalent Ion-sensing Receptor SCAR Is a Modulator of Gastric Secretion. J Biol Chem. 2001; 276, No. 43, 39549-39552.
95 Waisbren SJ, Geibel J, Boron WF and Modlin IM: Luminal perfusion of isolated gastric glands. Am J Physiol. 1994; 266 C1013-27.
96 Singh SK, Binder HJ, Geibel JP and Boron WF: An apical permeability barrier to NH3/NH4+ in isolated, perfused colonic crypts. Proc. Natl. Acad. Sci. U S A 1995; Vol. 92, 11573-11577.
97 Molecular Probes: BCECF Manual MP 01150, 2006 Updated: April 24, 2006; date of access: April, 8 2008. http://probes.invitrogen.com/media/pis/mp01150.pdf
98 Verkman AS: Development and biological applications of chloride-sensitive fluorescent indicators. Am J Physiol. 1990; 259, C375-88.
99 Verkman AS, Sellers MC, Chao AC, Leung T and Ketcham R: Synthesis and characterization of improved chloride-sensitive fluorescent indicators for biological applications. Anal Biochem. 1989; 178(2), 355-361.

100 Molecular Probes: Fluorescent Indicators for Chloride, Product Information. Updated: February 19, 2001; date of access: April, 8 2008. http://probes.invitrogen.com/media/pis/mp00440.pdf

101 Verkman AS, Sellers MC, Chao AC, Leung T and Ketcham R: Synthesis and characterization of improved chloride-sensitive fluorescent indicators for biological applications. Anal Biochem. 1989; 178(2), 355-361.

102 Molecular Probes: Online Handbook, Figure 21.17. date of access: April 8, 2008. http://probes.invitrogen.com/handbook/figures/0585.html

103 Biotium, Inc.: Product information. date of access: April 8, 2008. http://www.biotium.com/product/price_and_info.asp?item=52011&Sub_Section=02A

104 Gee KR, Brown KA, Chen WN, Bishop-Stewart J, Gray D and Johnson I: Chemical and physiological characterization of fluo-4 Ca2+-indicator dyes. Cell Calcium 2000; 27(2), 97-106.

105 Molecular Probes: Online literature. Fluo-4 Fluorescent Calcium Indicator. date of access: April 8, 2008. http://probes.invitrogen.com/servlets/publications?id=162

106 Molecular Probes: Online Product Information. Fluo Calcium Indicators. Updated: June 23, 2005; date of access: April, 8 2008. http://probes.invitrogen.com/media/pis/mp01240.pdf

107 Schnetkamp PPM, Li XB, Basu DK and Szerencsei RT: Regulation of Free Cytosolic Ca2+ Concentration in the Outer Segments of Bovine Retinal Rods by Na-Ca-K Exchange Measured with Fluo-3. J Biol Chem. 1991; 266, No 34, 22975-22982.

108 Molecular Probes: Online Handbook, Figure 19.30 Updated: April 5, 2005; date of access: April, 8 2008. http://probes.invitrogen.com/handbook/figures/0567.html

109 Molecular Probes: Online Handbook, Figure 19.20 Updated: April 5, 2005; date of access: April, 8 2008. http://probes.invitrogen.com/handbook/figures/0642.html

110 Hallows KR, Kobinger GP, Wilson JM, Witters LA and Foskett JK: Physiological modulation of CFTR activity by AMP-activated protein kinase in polarized T84 cells. Am J Cell Physiol. 2003; 284(5), C1297-308.

111 Hallows KR, Raghuram V, Kemp BE, Witters LA, Foskett JK: Inhibition of cystic fibrosis transmembrane conductance regulator by novel interaction with the metabolic sensor AMPactivated protein kinase. J Clin Invest. 2000; 105(12), 1711-21.

112 Mall M, Bleich M, Kuehr J, Brandis M, Greger R and Kunzelmann K: CFTR-mediated inhibition of amiloride sensitive sodium conductance by CFTR in human colon is defective in cystic fibrosis. Am J Physiol Gastrointest Liver Physiol. 1999; 277: G709–G716.

113 Clarke LL and Harline MC: CFTR is required for cAMP inhibition of intestinal Na1 absorption in a cystic fibrosis mouse model. Am J Physiol Gastrointest Liver Physiol. 1996; 270: G259–G267.

114 Grubb BR and Boucher RC: Enhanced colonic Na1 absorption in cystic fibrosis mice versus normal mice. Am J Physiol Gastrointest Liver Physiol. 1997; 272: G393–G400.

i want morebooks!

Buy your books fast and straightforward online - at one of world's fastest growing online book stores! Environmentally sound due to Print-on-Demand technologies.

Buy your books online at
www.get-morebooks.com

Kaufen Sie Ihre Bücher schnell und unkompliziert online – auf einer der am schnellsten wachsenden Buchhandelsplattformen weltweit! Dank Print-On-Demand umwelt- und ressourcenschonend produziert.

Bücher schneller online kaufen
www.morebooks.de

VDM Verlagsservicegesellschaft mbH
Heinrich-Böcking-Str. 6-8
D - 66121 Saarbrücken

Telefon: +49 681 3720 174
Telefax: +49 681 3720 1749

info@vdm-vsg.de
www.vdm-vsg.de

Printed by Books on Demand GmbH, Norderstedt / Germany